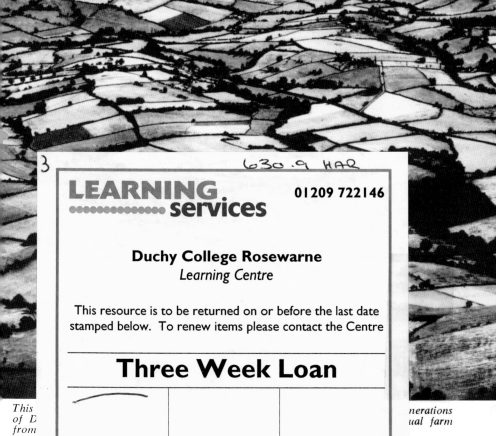

3

630.9 HAR

This .... *nerations*
of D .... *ual farm*
from

F .... ES

CONTENTS

Copyright © 1976 amd 1987 by Nigel Harvey. First published 1976; reprinted 1979. Second edition 1987; reprinted 2011. Shire Library 21. ISBN 978 0 85263 868 2.

Printed in China through Worldprint Ltd.

British Library Cataloguing in Publication Data available.

COVER: *Hundred Acre Field, at the foot of Beacon Hill, Ellesborough, Buckinghamshire.*

ACKNOWLEDGEMENTS
Illustrations are acknowledged as follows: Aerofilms, pages 1, 17, 21; Ashmolean Museum, pages 4, 7 (both), 10; Cambridge Antiquarian Society, page 13; Crown Copyright, Ministry of Defence (Air) from Cambridge University Collection, page 8; Farmers Weekly, pages 3, 11, 12, 15 (both), 18 (top), 20, 23 (all), 24, 25 (bottom right), 26, 27; Geoffrey Hammonds, page 25 (bottom left); Nigel Harvey, page 14; Cadbury Lamb, cover; Museum of English Rural Life, Reading, pages 19, 25 (top), 29; Peter J. Reynolds, pages 5, 6; Science Museum, London, page 2; Stewartry of Kirkcudbright Drystane Dyking Committee, page 30; J. A. Trant, page 18 (bottom); T. Weir, page 16. The maps on page 9 are based with permission on maps in *Mediaeval England: an aerial survey*, 1958, by M. W. Beresford and J. K. S. St Joseph. The drawings on page 28 were made by Mr P. D. Friend, with acknowledgement to Mr P. J. A. Lubbock of Firth Cleveland Ltd for information and to the *American Barbed Wire Journal*.

An open field village in the Middle Ages. The field in the foreground is the fallow field. As livestock are allowed to graze on the stubbles and weeds of this field, the two other fields, one carrying winter-sown corn, the other spring-sown corn, are protected by temporary fencing. This diorama is in the Science Museum.

This curving hedge now marks the boundary between the parishes of Cadbury and Stock-leigh Pomeroy in Devon. It originally marked the boundary of the estate of a Saxon called Cada and probably dates from the earliest days of the English settlement in this area in the seventh century.

THE MAKING OF THE FIELD

We all enjoy walking through fields, picnicking in fields, or leaning on a gate or sitting on a hill and looking at a landscape of fields. But we seldom pay much attention either to individual fields or to the field-systems of which they form part. Yet fields are among the most important creations of man. He has won them from forest and marsh, from estuary and upland; he has drained them and established their boundaries; he has adapted and re-planned them to meet changing needs. Generation by generation, he has lavished on them an infinity of thought, labour and skill. For the field is the home of the crops on which he and his livestock depend for food. Civilisation was made possible by the field and it still depends on the produce of the field.

So the fields we see around us when we walk or drive in the countryside tell us much about the past. They tell us about the life of our ancestors and the countryside in which they lived, about their tools and manner of working, about individual and collective effort, about the consequences in the farmlands of industrial developments in the distant towns, about the effects of economic and technical change. Thus,

3

The line of this Roman road in Hampshire is preserved by the field-system in which it forms so conspicuous a boundary.

the ridge-and-furrow pattern still found in clayland areas recalls the days before the invention and manufacture of the tile-pipe, when shaping the surface of the fields was the only way of draining them. Hedges and stone walls remind us of the dependence of our ancestors on local resources and on the skills and crafts they developed to exploit them, barbed wire of the contribution of the factory to the equipment of the farm. The shape and line of field boundaries may preserve the memory of a Roman road, the work of a medieval peasant, the plans of a Tudor landlord, an agreement between Stuart villagers, a Hanoverian Act of Parliament or the decisions of some Victorian magnate. Many purposes and many skills have gone into the making of the field.

It is the aim of this book to illustrate some of these purposes and skills so that the reader can better understand and interpret the creation and development of the quiet fields which still serve us.

4

*Ploughing with the primitive form of plough called an ard at the Iron Age farm recon-
structed by the Butser Ancient Farm Project at Butser Hill in Hampshire. The cattle are
Dexters, the smallest British breed, which are near in size to the original cattle.*

FIELDS OF FORMER TIMES

Men have cultivated the soil of Britain
since Neolithic times. The earliest
corn plots, worked by digging sticks,
have left few traces. The oldest visible
fields date from the Iron Age and sur-
vive only as small, square grass-grown
outlines in chalkland areas. They were
created by a light, primitive plough
called the *ard,* which, it has tradition-
ally been believed, could only scratch
the surface of the soil. Consequently it
was necessary to make a second row of
scratches at right angles to the first to
secure a proper tilth, which naturally
tends to produce a squarish field.
Recent trials with a reconstructed ard
at the Butser Ancient Farm Research
Project in Hampshire, however, suggest
that the ard can make quite an effec-

tive tilth by ploughing in one direction
only, so another explanation of the
shape of these fields may be necessary.

The oldest fields still under cultivation were formed by the *mouldboard
plough,* which we still use. This cuts
the soil vertically and horizontally and
overturns it by sliding it along a curved
surface, thus making a deeper and
better seedbed than is possible with the
ard. But it is heavier than the ard, it
needs more animals to pull it, and it is
more difficult for the ploughman to
lift and turn. Once the ploughman has
started a furrow, therefore, he is
anxious to keep his team and plough
on the move for a reasonable distance
instead of wasting time and effort by
stopping and turning unnecessarily. So

The 'business end' of a reconstructed ard, the primitive form of plough which broke but did not overturn the soil.

the new kind of plough created a new kind of field composed of a series of long strips, rectangular instead of square.

Such long, rectangular fields were the basis of a farming system which dates from Saxon times and in varied forms dominated much of the agricultural landscape for over a thousand years. Today it survives in only one parish in England, Laxton in Nottinghamshire, a unique fragment of an older countryside. This system is called the open-field system because the fields were not separated by hedges or other divisions but lay open and unenclosed in huge areas of ploughland.

We do not know how this system originated. But we can imagine a group of settlers in an empty countryside, who could at first muster only one ox-team, using this communal plough-team to till the land they were reclaiming from forest or scrub. In the autumn, they would use the communal plough-team day after day until rain, frost or snow stopped them, allotting each man the land they had ploughed in one day's work so that he could sow winter corn. Thus, if there were six men and thirty days' ploughing, each man would have five strips separated from each other by the strips of his neighbours. In the spring they would continue to plough elsewhere, allotting the strips as before, but this time sowing spring corn. But they knew that land gets tired if constantly cropped. So they added a third field for bare fallowing, which allowed the land to rest without a crop while stock grazed on the grass and weeds that grew there and provided manure for the soil. Here again each man had his separate strips.

So the open-field farm consisted of a number of small strip-fields, each representing one day's ploughing, scattered throughout three huge 'fields', with a rotation of winter corn, spring corn and bare fallow, and separated from neighbouring strips only by an open furrow. But the fields allowed a man to graze his stock as well as grow crops, for he could pasture his animals in them, either on their own when crops were not growing or else in the care of the village herdsman, who was generally one of the less able-bodied members of the community. Little Boy Blue, indeed, may have been too young for the job, for his neglect allowed sheep to graze in the meadow which was laid up for hay and the cows to trample the corn.

As time passed and population grew, each farmer had his own plough-team and the strip-fields advanced further and further into the surrounding waste-

ABOVE: *The outlines of Iron Age fields in Wiltshire. They were probably formed by cross-ploughing with a light, primitive plough.*

BELOW: *The open fields of Laxton, surrounded by modern enclosures. The open ploughland nearest the camera is the West Field; beyond it is the Mill Field, and to the left of this lies the South Field.*

The fields of Padbury, Buckinghamshire, showing the open field strips, preserved as ridge-and-furrow, on which the later system was imposed. The road shown runs between Hedges Farm and Grange Farm to the village just above the top of the photograph. See the opposite page.

land. But the principles of the system did not change.

Of course, there were endless local variations, changes and adaptations of this system to meet local needs and circumstances. There were also quite different systems in many parts of the country, some with hedged or embanked fields of the type we know today. Nevertheless, a visit to Laxton at harvest-time gives an unforgettable impression of the type of landscape which was the economic home of so many of our ancestors for so many centuries.

On one side of the village is the fallow field, a huge bare expanse of ploughland, where sheep graze on stubble and weeds. On another side lies the winter-sown field, some of the strips still carrying corn, some already harvested, some, where the farmer has been particularly ahead with his work, dotted with heaps of manure. On the third side lies the spring-sown field, more varied in cropping. Some of the strips are under corn, some under peas, vetches or clover, recalling the description of another such field by a traveller of the 1780s: 'it has no hedges, but every spot of it is uninterruptedly diversified with all kinds of crops of different yellows and greens which gives it a most pleasant aspect.'

Some of these open fields were enclosed, which means that they were divided into hedged fields, in the fifteenth and sixteenth centuries by landlords who converted ploughland into more profitable pasture. Others were enclosed in the following century and a half by private agreement between villagers. But most of them survived into later Hanoverian times, when the ancient system failed to meet the growing demand for food of the rapidly increasing urban population called into existence by the Industrial Revo-

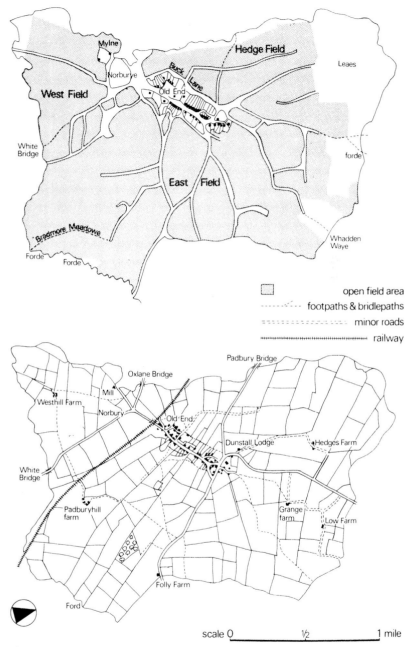

open field area
footpaths & bridlepaths
minor roads
railway

scale 0   ½   1 mile

The effects of Hanoverian enclosure. The parish of Padbury in Buckinghamshire from a map of 1591 (top), showing the three open fields, and (bottom) the parish after the enclosure in 1796. The land has been divided into individual farms, the farms into hedged fields.

9

lution. It was more adaptable than once believed and many enterprising individuals and communities introduced improvements and innovations. But it was difficult to grow new crops or establish new patterns of cropping when rotations were governed by the law and custom of the parish, while communal grazing prevented the improvement of livestock by selective breeding, since there was no way of preventing indiscriminate mating.

So change came in the drastic form of land redistribution called the Enclosure Movement which took place, sometimes by agreement, more often under the authority of local Acts of Parliament, in the later eighteenth and early nineteenth centuries. The old system was dissolved and the land reallocated to farmers who divided their new farms into new fields, dug ditches to drain them and planted hedges to give them control over the movement of their livestock.

The harsh social consequences of the Hanoverian enclosures, as of the earlier enclosures, are now part of history. But their physical consequences, the familiar countryside of compact farms and hedged fields, remain to this day.

Three ages of field-system in Oxfordshire. The small, squarish enclosures visible as crop-marks are Iron Age fields. The furrows continuing across the modern field from the allotments are the strips of a medieval open field. The modern hedged fields were formed by enclosure, probably in George III's time.

The Forty Foot Drain, one of the artificial rivers cut by Vermuyden in the seventeenth century to drain the Fenlands. The field-pattern is as carefully planned as the drainage scheme.

MODERN FIELDS AND THEIR STORIES

The fields of the modern countryside were formed in many different ways at many different times and they have many different stories to tell.

Many fields were originally won from forest or scrub, upland or marsh by pioneers now forgotten, perhaps a Saxon or medieval peasant, perhaps one of their descendants. The names of some of these early makers of fields are preserved in surviving records. One such, to take a random example, was Benedict, son of Siward, who in the early thirteenth century established his lonely farm at Colwich Town on Dartmoor, using the stones he cleared from the land which he brought under cultivation to make banks for the lane that still leads to the farmstead he

founded. But most of them have left no memorial save the fields they created.

Other fields were the work of great magnates who down the centuries have planned the extension and improvement of their estates. In the same period as Benedict was labouring alone on his uplands, the Earl of Warwick gave grants of land in the Forest of Arden to tenants who would reclaim and farm it. The Earl had many predecessors in such work. He also had many successors, for the tradition he represented continued to our time, when Lord Iveagh converted many hundred acres of Breckland into productive fields.

Indeed, the records of such men are

among our most valuable sources of information on our rural past. We know nothing except what we can deduce about Benedict's fields. But we know that the fields of the Earl's farmers were enclosed by a fence or hedge on a bank to protect the crops from the wild deer whose descendants the young Shakespeare later hunted.

The farmer, however, has won fields from the waters as well as from the wastelands. For example, the common Lancashire name of Rimmer is said to recall the peasants and fishermen who once lived on the rim of a huge lake near Ormskirk called Martin Mere. But you will not find Martin Mere on the map nowadays. It was emptied in 1783 by a canal-engineer who dug a channel four miles long from the lake to the sea, and the lake bottom has long been fertile farmland.

The most spectacular reclamation of this type was the drainage of the Fens, which converted a huge desolation into one of the richest agricultural areas in the kingdom. The story goes back to the Middle Ages. But the decisive period was the seventeenth century, when a company of investors under the Earl of Bedford employed a Dutch engineer named Vermuyden to ditch the southern Fenlands as a farmer ditches marshy ground. But he worked on a gigantic scale and his ditches were rivers, the largest, the Old Bedford River, being twenty-one miles long and seventy feet wide. By the 1650s the Bedford Level was carrying crops and stock, 'where never had been any before'. But as the reclaimed land dried and shrank and silt accumulated on the riverbeds, the rivers rose above the land they were supposed to drain. So hundreds of windmills were built to pump the water from the fields into the rivers. Yet still the rivers rose and the land fell and the Fenlands were not secured until the coming of steam power to drive the pumps. Thus, in the

This field was formerly mudflats on the edge of the Wash, which lies beyond the sea-wall in the background which turned this and other mudflats into fertile farmland. Now another seawall has been built and more mudflats in turn brought under cultivation.

Three generations of Fenland drainage pumps. The windmill probably dates from the eighteenth century. The steam plant was built in the 1840s. In the white building between them stands a diesel engine, installed in 1926.

words of an inscription on a Fenland bridge, 'water was conquered by her daughter, steam.'

The sea, too, has yielded a contribution, for the rich alluvial soil of tidal flats has inspired many 'innings', as schemes to recover land from tidal waters are called. One such scheme, incidentally, includes some of the handiwork of that most improbable of field-makers, the poet Shelley, who helped his friend William Madocks build embankments on Portmadoc estuary. But the most famous of all innings is that of Romney Marsh, once a shallow bay, now a level pastureland carrying more sheep to the acre than anywhere else on earth. This began in Roman or

13

pre-Roman times, was continued first by the monks of Canterbury and then, after the Reformation, by local landowners, and was finally completed in the 1830s. Among those who played a part in this centuries-long epic was Thomas à Becket. Today you can stand on the earthen wall where he once stood with a cloak over his finery, a map in his hand and his bailiffs around him. But where he saw shining mud, you see green fields, including a square mile which is still called the Innings of Saint Thomas.

At the other physical extreme, the fields of some hill areas represent the end of an early stage in agricultural development. For in former times farmers used to migrate for the summer months high into the hills so that their cattle could graze on the mountain pastures, while they themselves lived in a temporary house or improvised shelter called a *shieling* in the North of England and in Scotland, a *hafod* in Wales. Such summer migrations still continue in some mountainous parts of the mainland of Europe. But in this country some of the more remote 'summering-houses' have been abandoned and the others have been converted to permanent farmsteads and the wild grazings around them into fields. Thus in Wales many temporary upland shelters had become farms by the end of the seventeenth century, though the last summer migration took place as late as 1862. Many place-names still preserve the memory of this ancient system. Part of Beddgelert Forest in Wales, for instance, is called Hafod Fawr, meaning 'the big hafod', from a farm of that name which was once a true hafod, while Arecleoch Forest in Scotland is named after Arecleoch Farm, in the original Gaelic *airidh cleoch* or 'the shieling of the stones'.

Many fields, however, are comparatively recent in origin. For in large areas of Britain, particularly in the Midlands and the South, the great period of field formation or re-formation was between 1760 and 1830, when population increased at an unprecedented rate and drastic agricultural development was necessary to keep pace with the demand for food. In these years many new fields were won, notably from the open sheepruns of the chalk downs of the South and the wolds of

This causeway on Romney Marsh was once a high seawall built in the thirteenth century by Archbishop Boniface to reclaim mudflats from the sea by 'inning'. Later innings continued the work and the sea is now five miles away.

ABOVE: *One of the wooden culverts, made from a hollowed tree trunk, with one end open and the other fitted with a plug, formerly used by farmers on the Kent and Sussex marshes. It was placed between two drainage channels so that the water-level could be controlled by removing or replacing the plug.*

BELOW: *Field-making today, near Widecombe-in-the-Moor, Devon. The stones on the right have been removed from the field on the left by machinery and, since hedges do not grow on this poor, exposed upland soil, the new land is enclosed by factory-made fencing. Our ancestors cleared such land by hand and used the stones to build walls.*

ABOVE: *The shieling of Rowcheish in the Queen Elizabeth Forest Park in central Scotland. Originally a 'summering-house', it later became a shelter for hill shepherds.*
OPPOSITE: *An example from Norfolk of systematic enclosure. The order of the planned field system contrasts interestingly with the disorder of the unplanned village.*

the North, just as wasteland was brought under full cultivation during the Second World War. But many more new fields were created by the Enclosure Movement previously described.

The fields we see today, therefore, are the products of a long series of past achievements. They all once belonged to wild nature and were created, for the most part, by the labours of men working with axe and fire, mattock and spade, and only the power of their own muscles and those of their animals. They are relics of the days when 'all things were made by hand, and one at a time'. When we reflect on their story it is worth remembering that engineers reckon the strength of a man as one-tenth of a mechanical horse-power.

We should, however, remember that not all fields were created by farmers. Some were the by-products of other men's work. The builder and the ship-wright looked to the woodlands for their raw material and, as towns spread and ships grew larger, their demands increased. In the days before the exploitation of coal, too, the householder burnt wood on his hearth, the ironmaster and glassmaker burnt wood in their more rapacious furnaces. So, generation by generation, the forests shrank and the ploughman followed in the wake of the woodcutter. The formation of our fields, therefore, reflects industrial as well as agricultural history. So, as we shall see, does the story of their drainage and enclosure.

ABOVE: *By the end of the Middle Ages most of the workable land in the country had been reclaimed. But the high tide of cultivation later receded, leaving only the outlines of abandoned fields. Some of this 'marginal land', however, was again reclaimed during the Second World War. This picture shows fields of both types on the slopes of Dartmoor.*

BELOW: *A new field is made in a 1941 reclamation scheme in Powys. The ploughmen used the most advanced equipment then available, including heavy Lease-Lend tractors from the USA, and so became the first men to plough an acre of British soil in one hour.*

Water-meadows, which could be irrigated at will from nearby rivers by a system of sluices and channels and were thus able to produce a good crop of grass in dry seasons, were first made in Elizabethan times. A number were made in suitable areas at later times and some continued in use until only a few years ago. This photograph shows a 'drowner', the highly skilled worker who controlled the flow and level of the water, at work near Salisbury in the 1930s. These meadows probably date from the seventeenth century.

DRAINING THE FIELD

Farmers have always known that good drainage is essential to the production of good crops and from the earliest times they have planned their fields to take advantage of every means of ridding the soil of surplus water. In areas of poor natural drainage, for instance, our ancestors sited their fields so that they could plough up and down the slope of the land and the water could run freely down the furrows and into a ditch and away to a stream. But they frequently found it necessary to plough their fields into a series of parallel humps called ridge-and-furrow. In such fields the ridges are often two or three feet above the bottoms of the furrows, so that the land seems to be rolling towards you in waves, like the sea. This method of drainage by ploughing was practised until a century or so ago. It created, in effect, a series of small ditches that kept the ridge dry at the

cost of leaving the furrows in constant danger of waterlogging. But it was the best that could be done with the resources of the time.

Many ridge-and-furrows can still be seen, notably in the Midland clay areas, to bear witness to primitive methods of drainage and, since they were originally formed to grow corn but are now usually under grass, to changes in the farming system. They may also bear witness to changes in field-system, for in some cases a more recent pattern of fields has been imposed upon them. For example, the hedges of a Hanoverian enclosure may cut across the enduring ridge-and-furrow pattern of a medieval open field.

Most ridge-and-furrows are straight, but a number are curved in the form of an elongated and reversed S. These are relics of the medieval plough-teams of eight oxen yoked in pairs. Such

The weakness of the ridge-and-furrow system shown on Oxfordshire clay in a wet winter. The unridged field behind it has probably been drained with tile-pipes.

teams needed a good deal of space to turn at the end of a straight furrow, which meant much wasted or trampled headland, whereas only a narrow headland was required if the team ploughed a furrow which already partly turned them in the right direction. These S-shaped strips, however, record historical change as well as a historical technique. For they all date from the Middle Ages, whereas straight ridge-and-furrows continued to be formed for another three hundred years or so. This, it seems, reflects the better feeding of cattle in Tudor times onwards which enabled ploughmen to use fewer oxen in their teams and therefore to turn in a smaller space.

In the early nineteenth century, however, new needs made necessary new techniques. Population and the demand for food continued to grow, yet by this time there was little good land left to reclaim. Any substantial increase in production could only come from existing fields. One obvious method of improving crop yields was better drainage, which by lowering the level of water in the soil makes more plant food available to crops and allows earlier drying and warming of the soil in spring and therefore earlier cultivations and earlier growth. But the possibilities offered by traditional surface drainage were limited. In water-logged areas ditches sufficient to drain the land would have taken far too much space and divided the land into im-

The curved furrows ploughed by medieval ox-teams in the open fields of the Vale of Pickering in North Yorkshire have been 'fossilised' and preserved in the field-system created by enclosures. The S-shaped furrows were formed as the ploughmen manoeuvred their teams of eight oxen yoked in pairs for easier turning on the headlands.

practicably small fields, while ridge-and-furrow could only keep part of the soil dry. What was needed was some sort of 'underground ditch', a tunnel that would remove water yet allow men to work and cattle to graze over it.

The idea was not new. 'Hollow drainage', as it was called, had been used for many years in certain areas. Some farmers used twisted ropes of straw, brushwood, faggots or small branches buried in a trench to make such tunnels. Others used stones thrown in at random or built culverts of stone. But straw and wood decay and loose stones tend to get choked with silt. Stone culverts were better but they needed considerable skill to lay and were liable to collapse. The principle of hollow drainage was sound but an effective means of applying it was lacking.

The answer to this problem came not from traditional farm resources but from manufacturing industry. It first appeared at the end of the eighteenth century in the form of a curved tile, sometimes in the shape of the ridge-tile used on roofs, from which it takes its name, sometimes in the shape of a horseshoe. These tiles were often laid on flat tiles to prevent them sinking into the bottom of the trench, thus forming prefabricated versions of the old stone culverts. At first, they were made by hand and were therefore too expensive for many farmers. Later they were made by machine, though not always very successfully. A writer in 1844 reckoned that a stone drain was dearer than a tile drain but would last longer, as many tiles were badly laid or fell to pieces after a few years. Clearly in his time neither the makers nor the users of drainage tiles were masters of their trade.

Soon, however, the cylindrical pipe, called the tile-pipe after its predecessors, revolutionised agricultural drainage. Hand-made cylindrical pipes were known in the early nineteenth century, apparently invented by John Read, an ingenious Kentish gardener, who also invented a stomach-pump, a garden syringe and the familiar circular oast-house for drying hops. But they did not come into general use until the middle 1840s, when a method of mass-producing them, which greatly reduced their cost, was developed. Josiah Parkes, a drainage engineer, was one of the first to realise the importance of these cheap and effective pipes, which are still in use today. Holding one of them in his hand, he said to a landowner whom he knew: 'My Lord, with this pipe I shall drain all England.' He exaggerated, but there was some truth in his boast.

By 1880, thanks partly to government support, between two and three million acres, which was about one agricultural acre in twelve, had been deep-drained with the new pipes. Not all the work was well done, but huge areas lay drier and warmer and produced heavier crops at less cost. Many farmers of the time rightly called the pipes the most profitable crop they had ever planted.

The Victorian deep-drainage achievement was sufficiently important to leave its mark on contemporary fiction. Mr Mellot, a character in Charles Kingsley's *Two years ago,* refers to 'well-drained fields which ten years ago were poor clay pastures, fetlock deep in mire six months of the year', while Mr Brooke in *Middlemarch* advised another landowner to give draining tiles to his tenants, clearly regarding this as an obviously sensible investment. In another novel M. Yonge's *Hopes and fears,* a young squire compared the colour of his sweetheart's hair to that of ripe wheat, though he spoiled the effect by adding that he had harvested 'the best crop we have ever had from this land, a fair specimen of the effects of drainage'. She eventually preferred somebody else, which is perhaps not surprising, but he left on record the most engaging

In 1852 Wren Hoskyns published a wise and witty agricultural classic called 'Talpa or the Chronicles of a Clay Farm', which describes the rehabilitation of a semi-derelict farm in Warwickshire. His first improvement was drainage of the field with tile-pipes, for on this depended all other improvements. 'Talpa' is Latin for a mole, 'the fierce sub-navigator who taught us drainage'. The drawing is by Cruikshank.

" We shall learn of him another and a greater lesson, some day."

ABOVE: *Early nineteenth-century drainage tiles, showing their evolution from the simple horseshoe tile via the horseshoe on a sole to the cylindrical pipe. A modern tile-pipe stands in the centre.*

BELOW LEFT: *From 1784 to 1850 there was a tax on tiles, but an Act of Parliament of 1826 exempted tiles 'made solely for draining wet and marshy land', provided they were stamped 'drain'. The horseshoe tile in the middle stood on the flat sole held in the man's right hand. The tile at the top was made in an estate brickyard which recorded its annual output.*

BELOW RIGHT: *Some Victorian estates made their own drainage tiles from their own clay in their own brickyards. This early nineteenth-century horseshoe tile, found on the Dearing Estate in Kent, bears the estate's emblem, a fleur-de-lis.*

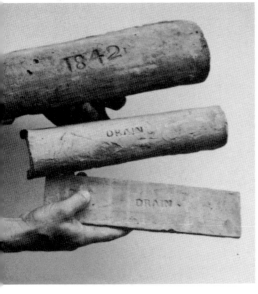

The drainer's tools. In the back row, from left to right, stand a topping spade and two bottom spades, in the front row a bottom scoop, a pipelayer and a drag, with tile-pipes in the background.

tribute ever paid to the work of the deep drainer.

During the long agricultural depression which began in the 1880s, however, drainage, like much else on farms, was neglected. No new work was undertaken and gradually the unmaintained ditches filled, the outfalls of the deep-drainage schemes choked and, as the flow of water ceased, silt collected in the pipes until they filled solid. A few schemes were maintained in working order but over much of the country the work of a generation of drainers perished.

The Second World War made necessary a new drainage campaign which continued in peacetime and from time to time the machines which have replaced the old drainage gangs turn up the remains of schemes laid a century or more ago. But the broken tiles and pipes of unfamiliar shape that you sometimes see in the spoil at the side of a field are not the only relics of this forgotten technical achievement. An enquiry some twenty years ago found that a number of mid-Victorian tile-pipe schemes serving several thousand acres had been reconditioned and were working well. So, more surprisingly, were several reconditioned stone drains, dating at the latest from early Victorian times and possibly from the eighteenth century.

ABOVE: *A Victorian mole-plough. Deep drainage by a pointed metal 'mole' (seen between the wheels), drawn through the soil to form an underground channel, is cheaper than tile drainage but less lasting. Mole-ploughs were originally pulled by windlasses turned by horses, then (as here) by steam power, and later by tractors.*

BELOW: *The outfall of a tile-pipe drainage scheme which was laid in 1864, reconditioned during the Second World War and is working well. But this is very exceptional. Most Victorian schemes were neglected in the long agricultural depression which began in the 1880s and have long ceased to flow.*

RIGHT: *Those who have watched machines laying tile-pipes will marvel at the achievement of the Victorians who deep-drained over two million acres by hand.*

ABOVE: *A well-laid Midland hedge. With proper maintenance, such hedges give good service indefinitely. Compare it with the neglected, gappy and patched hedge behind it.*
OPPOSITE: *A hedger at work laying and 'weaving' the bushes that form the hedge into a living, stockproof fence.*

ENCLOSING THE FIELD

There are many ways of enclosing a field. But for most of us, at least those of us who live in the lowlands, the hedge is the normal, almost the natural, way of doing so. For the hedge has dominated so much of the countryside for so long that we have come to regard it as an integral part of the landscape. So in a sense it is, but only because it does a job there. For essentially the hedge is no more than a piece of farm equipment, a device possibly for marking a boundary, certainly for restricting the movement of animals.

The farmer has used hedges for a very long time. Indeed, their characteristics tell us something about their his-

torical origins. They are living and growing, the product of a pre-industrial age when local needs were met by local resources provided by local nature, in this case, spiky bushes such as hawthorn. In recent years historians have developed ways of dating hedges by the number of species of plant they contain and have checked their results by comparison with old maps and other records. In general, a hedge will have one shrub species for every century of its life, so that a hedge of medieval origin may have an average of six or seven shrub species in a thirty-yard length, a hedge planted by the Hanoverian enclosers only two or three.

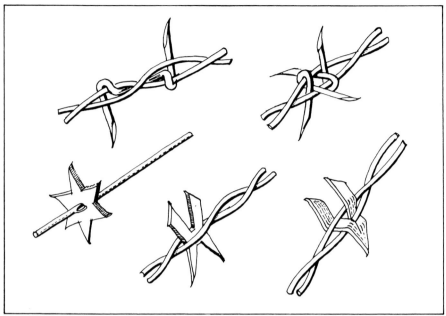

Early types of barbed wire. These were produced in America in the 1880s, the period when barbed wire was first introduced to Britain.

From this and other evidence we know that many hedges go back far beyond the period of the Enclosure Movement, some even to Saxon times. It is interesting to find that the word 'hedge-sparrow' was in use as long ago as 1530, and the word 'hedgehog' even earlier, in 1450.

Down the centuries hedges have served the farmer well. But they have their agricultural weaknesses. In particular, they require regular care and maintenance. They grow upwards, whereas the farmer wants them to grow sideways to create a stockproof barrier. Periodically, therefore, he has to 'lay' them by cutting into a proportion of the main stems that compose them, bending these cut stems over and fixing them into the ground with stakes, and then interweaving stems and stakes with withies to form a living wall which is later reinforced by new growth. All this takes time and skill. 'The billhook',

wrote Richard Jefferies a century ago, 'is the national weapon of the English farmworker', for the men of his day spent much of the winter using this tool to keep hedges in order.

For the hedge was not only important, it was also very conspicuous, and farmers prided themselves on their neat, well laid hedges which gave a favourable impression of their standards of farming workmanship. But it has been many years since many farmers could afford to maintain all their hedges properly. Most hedges were neglected during the long agricultural depression which began in late Victorian times and only a proportion have since been regularly cut and laid.

Hedges also take much space which can be used for growing crops. Above all, they are immobile and the positions in which they are sited .do not always meet the needs of later farmers. Even in early Victorian times, therefore,

A drystone waller at work building a mortarless wall, which, with proper maintenance, will last indefinitely.

some farmers were removing hedges in order to 'cut and carve ten or twelve large, square comely-looking fields out of thirty or forty unaccountably-shaped rhomboids', and a little later more hedges disappeared to make way for the working of the new steam-ploughs. But then came the agricultural depression and the farmer abandoned improvement. Instead he sought economy, and the factory was able to provide him with a means of keeping his field-boundaries stockproof at little cost.

This was barbed wire, an American invention developed mainly by the flamboyant John W. Gates, who wore three diamonds in his shirt front and combined good taste with a wholesaler's approach by buying Corots and Meissoniers by the square yard. Barbed wire reached Britain in the 1880s and by 1893 was sufficiently common to make necessary an Act of Parliament empowering local authorities to regulate its use. Soon it was widespread, sometimes forming new fences, sometimes nailed along hedges as a cheap, lasting and effective barrier which ended the need for expensive cutting and laying.

The field, however, still remained the same size. But the farmer may not want his livestock to roam all over a field. He may, for instance, want cattle to graze intensively, eating all the grass in one part of the field before being allowed to move on to another part,

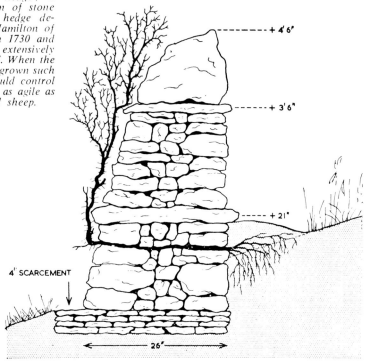

Section through a 'Galloway hedge', a combination of stone dyke and hedge devised by Hamilton of Baldoon in 1730 and later used extensively in Scotland. When the hedge was grown such a dyke would control even stock as agile as black-faced sheep.

+ 4' 6"

+ 3' 6"

+ 21"

4" SCARCEMENT

26'

which reduces wastage and encourages good grass growth. Such control is made possible by electric fencing, an easily movable barrier which enables the farmer to form temporary 'fields' within his permanent ones. This type of fencing was introduced from America in the 1930s, though it did not come into common use until after the war.

In recent years more and more farmers have removed an increasing number of hedges to create fields of a size which will allow the efficient working of modern machinery, save the cost of hiring a skilled man to lay them and eliminate a breeding ground for pests and weeds. This has caused a good deal of public concern as hedges not only add variety and a source of visual pleasure to the landscape, they also provide homes for representatives of most forms of wildlife found in Britain, mammals and birds, insects and plants,

most of them harmless, some of them helpful, all of them interesting. Some hedges, too, are of considerable historical interest. They are, indeed, ancient monuments with the same claims to preservation as other ancient monuments.

Fortunately, understanding between farmer, naturalist and conservationist is increasing and various field exercises involving representatives of these different interests have shown that compromises are often possible. Besides, when modern farming methods make the removal of a hedge necessary, the harmful consequences frequently can be countered by, for instance, the planting of small spinneys in an unploughed corner of a field.

The upland farmer, like his lowland brother, also used local materials to identify his boundaries and divide his fields. But his material was stone, not

30

bushes, and the story of the walls with which he patterned the hills is simpler than that of the hedges. They were built when the land was first reclaimed as part of the process of clearance and have since changed little. Climate and soil have ensured that man continues to farm the hills with livestock rather than machines and there has seldom been need to alter field-systems. But these miles of walling in remote parts, sometimes out of sight of human habitation and seen only by the occasional shepherd or hiker, bear witness to the mighty physical achievement of our ancestors.

They also illustrate the geological diversity of our islands and the skill with which the farmer has exploited it. The sandstone slabs of Caithness, the round boulders of Galloway, the millstone grit of northern England, the limestone of the Cotswolds, the granite of Dartmoor, all are very different types of material, each requiring a particular type of choice and arrangement from the wall-builder. If you watch such a builder at work, you will marvel at his mastery of his material. But remember that he is only one of the many craftsmen of the farm. It took a variety of skills and a variety of efforts by many men over many centuries to create the fields of the modern countryside.

FURTHER READING

Baker, A. R. H., and Butlin, R. A. (editors). *Studies of Field Systems in the British Isles.* Cambridge University Press, 1973.
Beresford, M. W., and St Joseph, J. K. S. *Mediaeval England; An Aerial Survey.* Cambridge University Press, second edition 1979.
British Trust for Conservation Volunteers. *Hedges.* 1975.
British Trust for Conservation Volunteers. *Dry Stone Walling.* 1977.
Chambers, J. D. *Laxton; The Last Open Field Village.* HMSO, 1964.
Darby, H. C. *The Draining of the Fens.* Cambridge University Press, second edition 1956.
Dodgshon, R. A. *The Origin of British Field Systems.* Academic Press, 1980.
Fowler, P. *Farms in England.* Royal Commission on Historical Monuments, HMSO, 1983.
Hall, D. *Medieval Fields.* Shire Publications, 1982.
Harvey, N. *Ditches, Dykes and Deep Drainage.* National Federation of Young Farmers Clubs, 1956.
Harvey, N. *The Industrial Archaeology of Farming in England and Wales.* Batsford, 1980.
Hoskins, W. G. *English Landscape; How to Read the Man-made Scenery of England.* BBC Publications, 1976.
Hoskins, W. G. *History from the Farm.* Faber, 1970.
Hoskins, W. G. *The Making of the English Countryside.* Hodder and Stoughton, 1977.
Millman, R. N. *The Making of the Scottish Landscape.* Batsford, 1975.
The Open University. *The Changing Countryside.* Croom Helm, 1985.
The Open University. *The Countryside Handbook.* Croom Helm, 1985.
Orwin, C. S. *The Open Fields.* Oxford University Press, third edition 1967.
Orwin, C. S., and Sellick, R. J. *The Reclamation of Exmoor.* David and Charles, 1970.
Parry, M. L., and Slater, T. R. *The Making of the Scottish Countryside.* Croom Helm, 1980.
Pollard, E., Hooper, M. D., and Moore, N. W. *Hedges.* Collins, 1974.
Rackham, O. *The History of the Countryside.* Dent, 1986.
Showell, R. *Hedges, Walls and Boundaries.* Dryad Press, 1986.
Taylor, C. *Fields in the British Landscape.* Dent, 1975.
White, J. T. *Hedgerow.* Dorling Kindersley, 1980.
Whitlock, R. *The English Farm.* Dent, 1983.

The '*Making of the Landscape*' series published by Hodder and Stoughton describes the historical and agricultural development of the landscape of individual counties.

PLACES TO VISIT

Acton Scott Historic Working Farm, Acton Scott, Church Stretton, Shropshire SY6 6QN. Telephone: 01694 781307. Website: www.actonscott.com/historic.php

Angus Folk Museum, Kirk Wynd Cottages, Glamis, Angus. Telephone: 01307 840288. Website: www.nts.org.uk/Property/5

Auchindrain Township Open Air Museum, Auchindrain, Iveraray, Argyll PA32 8XN. Telephone: 01499 500235. Website: www.auchindrainmuseum.org.uk

Beamish Museum, Beamish, County Durham DH9 0RG. Telephone: 0191 370 4000. Website: www.beamish.org.uk

Butser Ancient Farm, Chalton Lane, Chalton, Waterlooville, Hampshire PO8 0BG. Telephone: 0239 259 8838. Website: www.butserancientfarm.co.uk

Cogges Manor Farm Museum, Church Lane, Cogges, Witney, Oxfordshire. Telephone: 01993 772602. Website: www.cogges.org.uk

Fife Folk Museum, The Weigh House, High Street, Ceres, Cupar, Fife KY15 5NF. Telephone: 01334 828180. Website: www.fifefolkmuseum.org

The Great Barn Museum of Wiltshire Rural Life, High Street, Avebury, Marlborough, Wiltshire SN8 1RF. Telephone: 01672 539555.

Highland Folk Museum, Kingussie Road, Newtonmore PH20 1AY. Telephone: 01540 673551. Website: www.highlandfolk.com

Kent Life, Lock Lane, Sandling, Maidstone, Kent ME14 3AU. Telephone: 01622 763936. Website. www.kentlife.org.uk

Manor Farm Country Park, Brook Lane, Botley, Southampton, Hampshire SO23 8DH. Telephone: 01489 787055. Website: www3.hants.gov.uk/hampshire-countryside/manorfarm.htm

Museum of East Anglian Life, Stowmarket, Suffolk IP14 1DL. Telephone: 0844 5569245. Website: www.eastanglianlife.org.uk

Museum of English Rural Life, University of Reading, Redlands Road, Reading, Berkshire RG1 5EX. Telephone: 0118 378 8660. Website: www.reading.ac.uk/merl

Museum of Lakeland Life and Industry, Abbot Hall, Kendal, Cumbria LA9 5AL. Telephone: 01539 722464. Website: www.lakelandmuseum.org.uk

Museum of Lincolnshire Life, Burton Road, Lincoln LN1 3LY. Telephone: 01522 528448

Museum of Welsh Life, St Fagans, Cardiff CF5 6XB. Telephone: 0292 057 3500.

National Museum of Rural Life, National Museums Scotland, West Kittochside, Philipshill Road, East Kilbride G76 9HR. Telephone: 0131 247 4368. Website: www. nms.ac.uk

Norfolk Rural Life Museum, Beech House, Gressenhall, Dereham, Norfolk NR20 4DR. Telephone: 01362 860563.

Somerset Rural Life Museum, Abbey Farm, Chilkwell Street, Glastonbury, Somerset BA6 8DB. Telephone: 01458 831197.

Staffordshire County Museum, Shugborough, Milford, Staffordshire ST17 0XB. Telephone: 01889 881388. Website: www.nationaltrust.org.uk/main/w-shugboroughestate

Weald and Downland Open Air Museum, Singleton, Chichester, West Sussex PO18 0EU. Telephone: 01243 811363. Website: www.wealddown.co.uk

Yorkshire Museum of Farming, Murton Park, Murton Lane, York YO19 5UF. Telephone: 01904 489966. Website: www.murtonpark.co.uk